MARINE FISHES

Papua New Guinea, with its thousands of reefs and islands, supports one of the world's largest and most diverse marine habitats. Stretching as it does across one of the richest tropical areas of the southwest Pacific, Papua New Guinea is a meeting-place of many different seas. Marine animals and plants have converged and developed in the plankton-rich waters. Diversity is the key word. Marine life found in the southern waters adjacent to the Coral Sea and Australia's Great Barrier Reef, can differ considerably to that found in the Bismarck Sea area. However, marine fishes really know no barriers, and, provided conditions are suitable, will swim anywhere they can find a suitable habitat and food supply. Papua New Guinea shares most of its marine species with Indonesia, the Solomon Islands and northern Australia.

This does not mean these places are all the same. They most definitely are not. It simply means they are similar, sharing with each other a common pool of fishes, corals, molluscs, echinoderms, crustaceans etc. Plankton, that vast 'soup' of drifting life, has a great deal to do with the richness of different areas. It is basically the start of a food chain which ends as all food chains do, with the big carnivores.

In this booklet, I have endeavoured to present as simply as possible, a collection of marine fishes commonly seen by snorkellers and divers. I have also endeavoured to share in a small part, some of the information 34 years of diving has given me. There is far more to a fish than being served on a plate surrounded by fried potatoes. Each and every one of them is an individual. Their personalities differ not only from species to species, but from fish to fish. In a school, like a flock of sheep, they all look the same. However, spend a little time observing the school and you will start to notice certain individuals. There is the boss, the shy guy, the greedy, and so on. Once you start looking at the marine inhabitants this way, visiting the liquid world of innerspace becomes much more interesting.

Because of its tremendous diversity and lack of commercial exploitation, Papua New Guinea has a great deal to offer both snorkellers and divers alike. Already, overseas tourists are starting to visit on diving holidays. Usually they have to live on charter boats which cruise the isolated reefs and islands, stopping a day or two at the most interesting dive location. The Bismarck Sea, Louisiade Archipelago and Milne Bay areas are largely unexplored. With so much virgin territory, the visitor has the double excitement of not only an amazing variety of unafraid colourful fishes, but also exploring a terrain that possibly has never been seen by any human being before. Many new species of fish life are waiting to be discovered. Many more other types of animal and plant life as yet unknown live beneath Papua New Guinea's tropical waters.

It is a new area, a fascinating, wreck-studded playground for those with a sense of curiosity and adventure. The future is limitless provided exploitation is controlled. Over-fishing by dynamiting, monofilament nets, rubber-powered spearguns and the like, can decimate a reef area in weeks, destroying what could have been a very lucrative tourist attraction. Papua New Guinea is the most fascinating area in the world today. This is common knowledge. Its marine environment is equally as fascinating and is waiting there, in all its wondrous glory, for anyone with the incentive and sense of adventure to visit it.

1. The coastal and island-dwelling people of Papua New Guinea rely heavily on the sea as a source of much-needed protein. This young Trobriand Island boy from Kiriwina adds to the family food supply by spearing small fishes. Here he has a pennate fin butterfly-fish impaled on the end of his primitive spear. It will take several fish of this size to make a meal. It is difficult to explain to a young boy like this that overfishing will soon mean no fishing. Most reefs adjacent to civilisation have been overfished and the true beauty and richness of Papua New Guinea's marine world can only be seen on reefs away from extensive habitation.

2. White-tip reef sharks, *Triaenodon obesus*, are the most commonly seen around tropical reefs. Sleek, grey and graceful, curious and intelligent, they cruise the shallows, their pure white tips identifying them instantly. Rarely growing longer than 1.75m they are considered harmless though there have been two recorded attacks on spearfishermen who were handling struggling, bleeding fish.

3. Sharks have a never-ending fascination for people in general and divers or snorkellers in particular. Fortunately most species in Papua New Guinea waters are not potentially dangerous. Spearfishing does sometimes cause frightening confrontations but normal non-spearfishing divers and snorkellers have really nothing to fear. The black-tip reef shark, *Carcharhinus limbatus*, in spite of his rather aggressive appearance, poses no threat to the casual observer. They rarely grow longer than 2m.

4. The Greater Hammerhead shark, *Sphyrna mokarran*. The strangely-shaped head with eyes so far apart gives this shark tremendous vision. In spite of its fearsome appearance, the hammerhead appears to be a gentle, unaggressive creature. There are no recorded attacks on humans by hammerhead sharks in Papua New Guinea waters.

1

3

2

4

5

6

7

5. The friendly and harmless, remora or sucker fish, *Echineus naucrates*, has a type of suction cap on its head enabling it to 'hitch a ride' on any large fish or shark. When sharks are feeding, their remoras will leave the host and join in the feast. Appearing to be opportunists they do, however, earn part of their keep by feeding on their host's parasites. Because they possess a triangular dorsal fin the layman will often mistake remora for sharks. Fortunately they are harmless and probably because of their trust in large creatures will sometimes attach themselves to a diver. They have no teeth which is just as well for they are greedy feeders and will often suck in a finger or two with any proffered food.

6. This agate-eyed moray eel, *Gymmothorax flavomarginatus*, is being cleaned by the busy little *Labroides dimidiatus*. A friendlier creature would be hard to find and like all his family, he loves a scratch and free food handout. The opening and shutting of the mouth, which can look so threatening, is merely their method of breathing. They will never attack but if cornered and hurt, could bite in self defence. Nocturnal hunters, morays spend most of the daylight hours peering from some coral cave or cranny. They can grow up to 3m long.

7. Camouflage is the only defence for the winged pipe fish, *Halicampus macrorhynchus*. They blend into the habitat so well that a diver will not see them. They appear to peck small particles from weeds but exactly what they are eating is hard to define. This specimen is about 20cm long. They are listed as very rare but this could be because they are so difficult to see.

8. Unfortunately this lovely fish is rarely seen by observers from the surface. The six-banded rock cod, *Cephalopholis sexmaculatus*, is rather shy, preferring the confines of caves and overhangs to swimming in the open. A good table fish, island people sometimes spear them for food. A carnivore, it is possibly a nocturnal feeder; its secretive daytime habits certainly suggest this.

9. The curious cod, *Epinephelus microdon*, is not a true cod, but belongs to the grouper family. The curious cod is one fish who should be protected. They have no fear of humans and generally will swim straight over for a better look. This friendly attitude has resulted in their being extremely rare; a fat juicy cod who swims up to a spearfisherman has only one fate. His friendly personality makes him a great companion on a dive. They are carnivores and will eat from your hand without fear. This specimen weighs about 2.5kg.

10. A popular eating fish, the coral trout, *Plectropoma leopardus*, is commonly found throughout the tropical Indo Pacific. It is a ferocious carnivore forever on the lookout for unwary little fish. Adults reach up to 70cm in length. The little labroides cleaner wrasse is seen here cleaning its predator's gills with complete safety.

11. Imitation is common among the marine fishes. Usually a non-venomous fish will imitate a venomous fish as a form of protection. The false cleaner or sabre tooth blenny, *Runula rhinorhynchos*, is an exception. It uses its similarity to the cleaner wrasse as a type of camouflage enabling it to swim close to larger fish, and instead of nipping off parasites, actually nips out a piece of flesh and escapes unscathed.

12. Nowhere on land will wild creatures approach man as they do in the water; the jack, *Caranx luabris*, can be so curious and friendly that he becomes a nuisance. They range in colour from almost black to a dull, deep silver and will readily eat from a diver's hand.

13. The blue-fin trevally, *Caranx melampygusy*, is similar in many ways to *Caranx luabris*, except he prefers to swim with a companion or two. The blue fin is also more commonly seen. Like all members of its family it is a savage hunter, cutting through schools of small fish with deadly speed.

9

10

11

12

13

14

15

16

17

18

14. Unlike reef fish who usually remain within an easily defined territory, school fish, moving as they do, like great flocks of fast-flying birds, appear to be the gypsies of the sea, with no definite home area. Of course some species such as the marlin and giant tuna are true ocean roamers, but most school fish do have a territory. Sometimes it may stretch hundreds of kilometres but frequently, as in the case of these jacks, *Caranx sexfasciatus*, compared to the large number of fish, their territory is fairly small. If one is fortunate enough to meet a school of jack, stay still and they will crowd around in a living wall of curious, quick-moving fish. A fast, mysterious drumming can be heard as the tails flick and the fish turn. It can be one of life's finer experiences, but unfortunately one of life's fast disappearing ones.

15. Red Bass, *Lutjanus bohar*, are often called snapper and possess a personality as mean as their appearance. Vicious, fast-moving predators, they compete with sharks for food without fear or respect. They grow up to 3kg in weight and are considered poisonous in Australia.

16. The word 'snapper' conjures up a vision of a large variety of good, edible sea fishes. The largest, most popular of the tropical snappers is the red emperor, *Lutjanus sebae*. This specimen is a juvenile around 1.5kg; the lovely red stripes will fade into an overall pink as the fish matures.

17. Not as popular as the red emperor, but still good eating, are the yellow-finned snapper, *Lutjanus fulvus*, which grows to around 1.5kg in weight. It is a carnivore and prefers a solitary life.

18. The four-lined snapper, *Lutjanus kasmira*, however, is rarely seen alone. These lovely fishes are fully grown when about 30cm long. They are gentle fish who generally swim in small schools, moving over coral growths like a living patch of sunshine; a delight to all who are lucky enough to see them.

19. A most popular creature is the friendly cave-dwelling soldier fish, *Holocentrus spinifer.* Although night hunters, they are easily seen during the day sitting in front of their cave. They are often speared by night fishermen which indicates they are good to eat.

20. The red-orange squirrel fish, *Mypristis vittatus*, is also a cave-dweller who hunts under cover of darkness, but prefers to live in a close-knit family group. Whereas the soldier fish prefer to hunt small fish, squirrel fish eat tiny planktonic creatures and are most prolific in reef areas where big currents are prevalent. The large black eye is another indication that these fish are night predators.

21. & 22. The lunar-tailed glass eye, *Priacanthus hamrur*, is definitely a personality fish and seems fascinated by a diver's antics. Instant colour change is common and these two photos show the same fish at different times. As the fish did not swim into his cave when he turned paler, fear was not the cause. It was probably a change of attitude towards the diver that brought out the pale stripes. Many different fish are capable of instant colour change.

23. Surrounded by tiny jewel fish, the Goldman's sweetlips, *Plectorhynchus goldmani*, stands out as a true reef beauty. They are a great favourite with snorkellers because they often inhabit shallow water and provided no one has tried to spear them, generally do not swim away when approached. They grow up to 60cm long.

19

20

21

22

23

24

25

26

27

28

24. The striped sweetlips, *Plectorhynchus cuvieri*, is not commonly seen. They are not as easily approached as some members of the sweetlips family, and are often confused with Goldman's sweetlips. An easy way to pick the difference is to look for the heavily-spotted tail that is particular to Goldman's sweetlips. Also, their body is more heavily striped, the black bars almost unbroken along the full body length.

25. Many-spotted sweetlips, *Plectorhynchus shaetodontoides*, is a good descriptive name for this lovely fish. A cleaner wrasse is seen protruding from the sweetlips' gills, busily removing parasites from the gill rakers. It grows to about 55cm.

26. Because of their vivid colouring, even small fish instantly attract the eye. The orange fairy basslet, *Anthias squammipinnis*, is one such fish. They live in large schools hovering above the coral in a shimmering cloud of colour. When danger threatens they immediately flit into the coral's sheltering arms.

27. Brightly-coloured jewel fish, *Anthias sp.*, photographed on a huge solitary reef, fill their blue-green world with a kaleidoscope of colour and movement. Like so many of the fish family, basslets or jewel fish have a male with a harem of many females. There is always a dominant female who will replace the male if he should be removed.

28. There seems to be no common name for this sand- dwelling fish, *Parapercis tetracantha*. Perhaps banded sand fish is easier to remember. Small fish were offered to coax him into position for this photo and he treated other family members harshly when they tried to steal a mouthful.

29. The masked monocle bream, *Scolopsis billineatus*, is a funny little fish whose curious nature attracts him to closely observe the bizarre and unusual. This makes it a very easy fish to watch. They move in short, sharp darts here and there, but never far from the safety of a cave or coral hole. This habit is common to all members of the monocle bream family. They are easily photographed and generally happy to accept a handout of raw fish particles.

30. Unlike the masked monocle bream who lives a solitary life, the large slate bream, *Spilotichthus pictus*, loves to be the centre of a crowd. These fish prefer to inhabit areas close to cleaner wrasse stations, hovering over a coral bommie or clump, in a dense cloud. They are plankton eaters. The juvenile of this species is brightly-striped and sometimes called the rainbow bream.

31. Fusiliers grow up to 30cm in length. There are many species most of which are beautifully-coloured with yellow, blue or brown stripes. The black-tipped fusilier, *Caesio chryozonus*, is very common on most coral reefs. They are fast swimmers dashing in glittering schools along the reef edges. They like sunlight and appear to feed on shallow-water plankton which makes observation of this fish family relatively easy. Like many small school fishes they fall prey to most of the larger carnivores including coral trout, tuna, turrem and sharks.

32. This species of fusilier, *Paracaesio pedleyi*, swims in huge, fast-moving schools along the reef edges. They are mainly plankton feeders but are not adverse to a little mashed fish if it is offered. They are a common sight throughout the south-west Indo Pacific. This photo shows a small section of a school that consisted of thousands of fish. They swam past continuously for 30 minutes or more, a most wonderful sight, their blue bodies against the darker water.

29

30

31

32

33

34

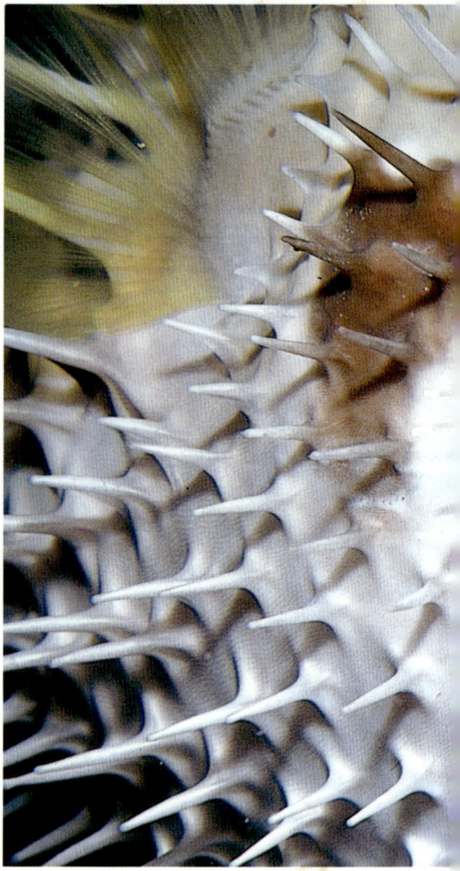
35

36

33. This snobbish-looking fellow is called the scribbled puffer fish, *Arothron mappa*. They can be deadly poisonous to eat if not prepared correctly and this is best left to Japanese chefs. When danger threatens they inflate their bodies with water, hopefully making themselves too large to swallow. Their very abrasive skin which takes the place of scales also acts as a deterrent to predators. They like to eat a variety of different foods such as shells, echinoderms, ascidians and seaweeds. This specimen is quite large and about 2kg in weight.

34. A close relative to the scribble puffer fish is the white-spotted puffer. They are slow swimmers, cruising along like bewildered blimps, their fins waggling furiously. They have a very strong bottom tooth, hard enough to scrape shells from rocks or dead coral. A quiet approach is necessary so as not to cause alarm, but as always when in the marine world, once the fish feels you are harmless and pose no threat, it goes about its daily business without any apparent fear or nervousness. The scientific name for this amazing-looking fish is a mystery. Although one may exist, it proved impossible to find before publication.

35. The French porcupine fish, *Diodon holacanthus*, can grow just as large as his cousin, the scribbled puffer. An amazing attribute of the porcupine fish is his ability to puff into a round spine-covered ball when danger threatens. It is quite easy to make this happen, simply by pulling his tail; a more unattractive mouthful would be hard to imagine. When inflated they should be handled with care.

36. Quite the most magnificent fish in the ocean, the clown triggerfish, *Balistoides conspicillus*, is much sought after by aquarists. Unfortunately they fetch a high price and their popularity has caused a tremendous decline in numbers. They have skin instead of scales and appear to lead fairly solitary lives. Their flesh is thought to be poisonous.

37. One of the largest members of the triggerfish family, the orange-faced triggerfish, *Pseudobalistes flavimarinatus*, has a personality to match his looks - strong and individual. They live in territories adjacent to one another but jealously guard their chosen area and woe betide any unwanted intruders. When hungry they are quite fearless in their search for food; brittle stars, sea urchins and small scallops can all fall prey to a foraging oranged-faced trigger.

38. A night photograph of the unicorn surgeon fish, *Naso brevirostris*, with its marvellous elongated horn. Like all surgeon fish, they have rough abrasive skin which is very popular with Trobriand Island woodcarvers for use as sandpaper. This species is considered fairly rare, however, they, like so many other fish not commonly found elsewhere, abound around the offshore reefs of Papua New Guinea.

39. Not commonly seen, but worth looking for, blue tang, *Paracanthurus hepatus*, are shy in the extreme. Looking more like bright paper cutouts than real fish, they are often found harmoniously sharing the same coral clump as a family of dascyllus. This type of cohabitation can generally only be successful when each group has different feeding habits though in this case, both species appear to be feeding on plankton. They are popular with aquarists but unfortunately do not survive well in captivity.

40. The white-cheeked surgeon, *Acanthurus glaucopareinis*, displays his scalpel-like appendage with a bright yellow warning flash. This razor-sharp spine on the caudal peduncle is used for defence and to ward of unwanted intruders.

41. The black-edged *Dascyllius reticulatus*, is extremely common and belongs to the damsel fish family which includes all the anemone fishes. They are frequently seen in large schools drifting like living jewels over the corals.

37

38

39

40

41

42

43

44

45

42. Another view of the black-edged *Dascyllius reticulatus*, which illustrates how versatile they can be in their selection of coral habitat. Like the anemone fish, these dascyllius also love to be offered a few fish scraps. However, unlike the anemone fish who feed their anemones, dascyllius eats the food himself. They grow to about 3cm in length and are extremely common.

43. The lemon damsel fish, *Amblyglyphidodon aureus*, lives a solitary life. Not for him the crowded coral tenements of his cousins. He prefers a small and diverse area which he constantly patrols. All damsel fish are fairly small, rarely exceeding 5 or 6cm in length.

44. It was late afternoon and the reef fish were settling into the corals for the night. *Amphiprion clarkii*, pictured, is probably sound asleep; certainly she never moved when this picture was taken. Anemone fishes are fortunate in that they do not have to hunt for a safe bed but simply snuggle into the protecting tentacles of their anemones.

45. Another picture of the anemone fish, *Amphiprion clarkii*, in a completely different type of anemone and with different colouring. This colour difference is quite common among young anemone fish and can cause some confusion. *Clarkii* has been seen and photographed without stripes at all, but this is rare. Most anemone fish do conform to a pattern. Usually only the fishes colours may be brighter or darker.

46. Anemone fish, or clown fish as they are often called, are everyone's favourites. Easily observed because of their reluctance to leave the static host anemone, they can be watched by a snorkeller for hours on end without very much effort. The pink anemone fish, *Amphiprion perideraion*, is one of the most common species. Just why the anemone host occasionally balloons in this way is not clear.

47. Panda clown fish, *Amphiprion polymnus*; in 35 years of diving I have only seen this species once. The host anemone is relatively small compared to that of their cousins. The orange patch on the sand is actually a clutch of eggs laid on a dead sand dollar.

48. *Amphiprion percula* is perhaps the most colourful of the clown fish. Although not as common as the pink anemone fish, most reef areas support several colonies of these beautiful little creatures. *Percula* has a distinctive undulating swimming motion which makes it extremely noticeable and rather appealing.

49. Flashlight fish, *Photoblepharon palpebratus*. Underwater, they appear to be a galaxy of winking and moving stars. The brilliant light is provided by thousands of luminous bacteria that live in a light organ under the fishes' eyes. A special upwards-blinking eyelid gives the winking effect. Winking or blinking can occur at a rate of 75 blinks per second which gives the effect of a marine solar system rather than a school of little fish. The lights are thought to be used for hunting the small planktonic animals on which these fish live, as well as for individual identification. Any disturbance and all the lights will disappear, flicking back on when calm returns.

50. Flashlight fish with the light extinguished.

51. The common names of fish often change from place to place. This magnificent angelfish has several common names: the imperial angelfish, the masked angelfish and the emperor angelfish. All the names suit him very well, but his scientific title, *Pomacathus imperator*, is the same everywhere. Owing to a shy personality he is not often seen.

47

48

49

50

51

52

53

54

55

52. The blue angelfish, *Pomacanthus semicirculatus*, lives a fairly solitary life. Surprisingly the young appear to be almost identical to that of the imperial angelfish. That the adults are so different is one of nature's mysteries. Blue angels can become very friendly and love to have a dead coral plate turned over so they can feed on the delicate sponges beneath, however, their greatest love is oysters which they rely on other oyster-eating fishes to open.

53. The lovely little beaked coral fish, *Chelmon rostratus*, is certainly a delight to see. His manner of feeding is similar to that of the long-nosed butterfly fish. The false eye spot on this fish's dorsal fin is probably to confuse potential predators by making its head appear to be at the rear of the body. His real eye is well camouflaged. This confusing eye spot pattern is common to many species of butterfly fish.

54. The long-nosed butterfly fish, *Fordcipiger longirostris*, with its tiny mouth on the end of a long nose is indeed an unusual-looking fellow. Like most butterfly fish they generally swim in pairs moving in unison across the reef corals. The long nose is used for reaching into deep rocks and crannies in search of food which probably consists of small worms, fish eggs and the like.

55. A butterfly fish whose territorial habits are similar to those of the black-backed variety is the pyramid butterfly fish, *Hemitaurichthys polylepis*. They swim in large schools numbering up to 50 or more. Often fishes of other species will join these schools and to see them swimming together is a grand and colourful sight. Prolonged observation will show an amazing coordination as they swim together then stop to feed all at the one time.

56. Often seen swimming in schools or groups of more than two, the black-backed butterfly fish, *Chaetodon melannotus*, ranges across a large area of reef. This is somewhat different to that of most butterfly fish who prefer a smaller territory which they often guard diligently.

57. Sometimes called the zebra butterfly fish, *Chaetodon pelewensis*, is noticeably smaller than other members of its family. Swimming in pairs, they flutter in unison from coral to coral, pecking tiny food particles from small algae-covered caves and crannies. Being small in size, their true beauty is only appreciated at close range.

58. Easily identified by his black mask and white headband, the raccoon butterfly fish, *Chaetodon lunula*, is common on all Papua New Guinea reefs. It is somewhat larger and leads a mainly solitary life, characteristics not common to most other members of his family.

59. The ornate butterfly fish, *Chaetodon ornatissimus*, usually swims with one or two companions and seems to prefer shallower waters down to perhaps 20 metres. Like all their family they prefer areas of rich, hard coral growth which offers many holes and caves into which they can dart if danger threatens.

60. Perhaps the most elegant of all fishes, the Moorish Idol, *Zanclus cornutus*, is in a family all its own. Usually seen in pairs they are common inhabitants of all Indo Pacific reefs. If one Moorish Idol is separated from its companion, there is mild panic until they are reunited. Their loyalty is a touching thing to see.

61. Often mistaken for the Moorish Idol, the banner fish, *Heniochus diphreutes*, is similar only in colour and shape. Its personality is much more definite. They tend to feed in mid water, catching tiny planktonic animals as they drift past in the current. This one is the more rare of the two known types.

56

57

58

59

60

61

62

63

64

65

66

62. The masked banner fish, *Heniochus monoceros*, has a completely different lifestyle to his cousin. Somewhat larger in size than most of his family, the masked banner fish prefers to inhabit caves and overhangs rather than the open water. They appear to live a fairly solitary life. The horny point on their forehead develops with age.

63. & 64. Parrotfish, with their bright colours and large size, are among the most visible of the reef inhabitants. This lovely blue-barred fish, *Scarus ghobban*, is either a female or an immature male. The beak- like teeth are easily seen; not a true beak but it appears to be many small teeth fused together. The design is perfect for they graze on algae growing on dead coral, removing the vegetable matter and excreting the remainder as fine white sand. The second specimen pictured is asleep. Because they lack eyelids, sleeping fish appear to be wide awake, however, their brain is asleep.

65. Although many parrotfish species live in schools or family groups, the long-nosed parrotfish, *Hipposcarus longiceps*, seems to be a loner. Pictured here while sleeping at night, his true blue colouration has been subdued by a mottling of his scales. Many fish change colour at night, probably as a form of camouflage. Most parrotfishes have at least one breeding season before starting to change sex. This is why the same species can look so different in colour. Years ago these strange sex habits were unknown to science. Male and female parrotfish were often described as different species because of this. Scientists are still studying this phenomenon of sex-changing so common among marine fishes.

66. The wrasse family is the largest of all the fish groups. They are usually bright-coloured and surprisingly intelligent. The blue tusk fish, *Choerodon schoenleinii*, digs for food in the sand. His busy excavating uncovers worms, shrimps, shells and probably fish eggs; all of which the tusk fish quickly snaps up.

67. This male moon wrasse, *Thalassoma lutescens*, is head of a harem of females; these are his wives. He protects them against other amorous males with great vigour. If this male should die or fall prey to an enemy, the strongest of his wives will change sex and eventually take over the duties of her former husband.

68. The unusually-shaped mouth of the bird wrasse, *Gomphosus varius*, distinguishes them from all other wrasses. Crabs and shrimps hiding in holes and small cracks, are easily caught. This female bears little colour resemblance to the male which is dark blue and turquoise in colour and somewhat larger.

69. Manta rays, *Manta alfredi*, are large, gentle plankton feeders. To meet one underwater is a privilege. Unlike most rays, they fly through the ocean on their great triangular wings, sieving food particles directly from the water. Their young are born live and fully formed, ready to fend for themselves in the open sea.

70. In spite of their sluggish ways, scorpion fish, *Scorpaenopsis venosa*, have little to fear. They possess a row of long, needle-sharp, toxin-coated spines along their backs. Although not deadly to humans, the toxin can cause extreme pain and swelling. It is wise to leave all species of the scorpion fish alone. Looking more like a piece of the reef than a fish, their camouflage plays a large part in the daily hunt for food as most of the Scorpadie family prefer ambush to pursuit when it comes to hunting.

71. The horrid stonefish, *Synanceia horrida*, a close relative of the scorpion fish, is extremely venomous. Their camouflage is excellent, it being quite easy to tread on them as they do not swim away when approached, preferring to rely on the potent venom in their dorsal spines as a deterrent to potential danger. They are most dangerous when walking on reef areas in shallow water.

67

68

69

70

71

72. *Pterois antennata*, a most beautiful member of the scorpion fish family, has many different names; lion fish or zebra fish seem to be the most common. Unlike their static cousins, they float above the corals like marine butterflies, silky pennants fluttering from their long toxin-coated spines. They are often found in family groups. A more beautiful sight is hard to imagine than these exquisite fishes flying together over a garden of corals. They are easily approached, relying on their venom for protection. Like many members of their family, they are nocturnal hunters, sneaking around under the cover of darkness in search of small sleeping fish. They are common to all Pacific reefs but in Papua New Guinea waters, are far more prolific than anywhere else.

73. Batfish, *Platax teira*, are commonly seen around most coral reef areas. They are slow-moving plankton feeders who sometimes swim in quite large schools. This loner has made his home under the crow's nest of a sunken Japanese warship in Rabaul Harbour. The masts on a ship, with their protruding spars and stays, make a fine home for *platax* and they are common wreckage inhabitants, treating the growth-covered man- made construction as a true reef area which, if you think about it, is exactly what it has become. Papua New Guinea has a fine collection of marine wrecks, most of which support a tremendous number of fish species. Many are in shallow water and offer excellent fish-watching to the snorkeller. Batfishes are seen on and around most Papua New Guinea wrecks.

74. The dragon flathead, *Platycephalus sp*, is a most amazing creature. His personality is very similar to that of the scorpidae family and his dorsal spines are coated with a nasty toxin. Because of these similarities, dragon flathead are often confused with scorpion fishes. They can grow quite large, up to 1.5m long and are excellent eating. However to kill such a wonderful creature would be a great pity for they are fairly rare and because of their docile habits, easily speared.

72

73

74